Number CONNECTIONS

second edition

Pupil Book

Purple Level

ROSE GRIFFITHS

Heinemann Educational Publishers
Halley Court, Jordan Hill, Oxford, OX2 8EJ
a division of Harcourt Education Ltd
www.myprimary.co.uk

Heinemann is a registered trademark of Harcourt Education Ltd

First edition first published 1996

Second edition first published 2005

10   09   08   07
10 9 8 7 6 5 4 3 2

ISBN 978 0 435 02172 6

Designed and typeset by Susan Clarke
Illustrated by Mick Reid and Jeff Edwards
Cover design by Susan Clarke
Printed and bound in Scotland by Scotprint

The author and publishers would like to thank teachers at the
following schools for their help in trialling these materials:
Humberstone Junior School, Leicester
Knighton Fields Primary School, Leicester
St Anne's Primary School, Streetly
Orton Wistow Primary School, Peterborough
Spooner Row Primary School, Wymondham
St John's Primary School, Sevenoaks

Contents

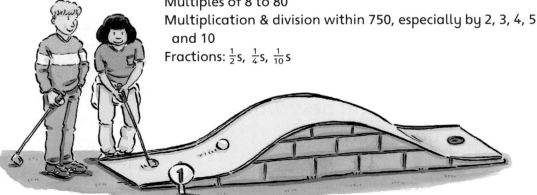

Using this book

Your teacher will talk to you about where you will start in *Number Connections*.

  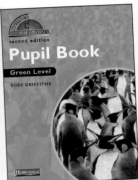  

Getting started

Check that you <u>can</u> do the first two pages in each part of this book, before you do any more.

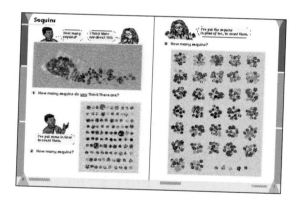

Reading

There are word lists in the *Teacher's Guide*.

These will help you learn any new words you need.

I've made cards from my list.

Extra activities

There are more activities and games in the *Copymasters* and the *Games Pack*.

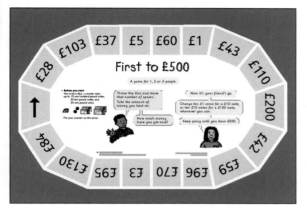

Take them home for extra practice!

We like doing the SpeedyTables.

Can we get more right, <u>and</u> get quicker?

Progress tests and Record sheets

These are in the *Teacher's Guide*.

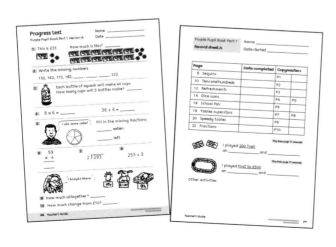

Check on your progress ...

and keep a record of what you've done!

Part 1
Contents

Counting and place value
Addition and subtraction
Multiplication and division
Mixed problems

Sequins

How many sequins?

I think there are about 100.

1 How many sequins do <u>you</u> think there are?

I've put some in lines to count them.

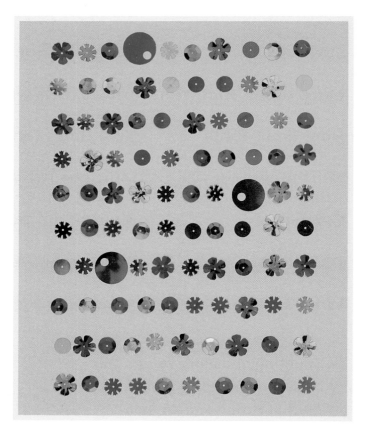

2 How many sequins?

I've put the sequins in piles of ten, to count them.

3 How many sequins?

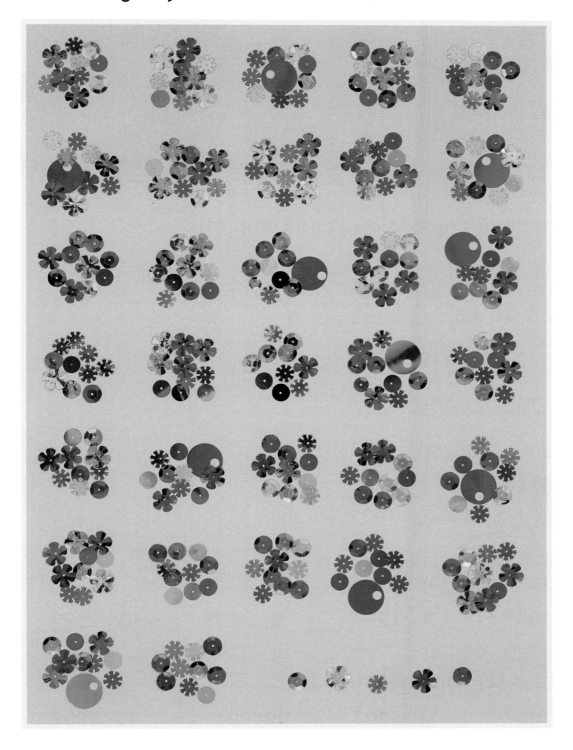

Tens and hundreds

I use my tape measure as a number line. It goes up to 300.

170, 180, 190, ...

The red numbers are multiples of ten.

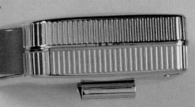

178 179 180 181 182 183 184 185 186 187 188 189 190 191 1

What comes next? Copy and complete.

1 60, 70, 80, _____, _____, _____.

2 80, 70, 60, _____, _____, _____.

3 120, 130, 140, _____, _____, _____.

4 170, 180, 190, _____, _____, _____.

5 300, 290, 280, _____, _____, _____.

100 is one more than 99.

What is one more than:

6 100?   **7** 199?   **8** 200?   **9** 299?

What is ten more than:

10 100?   **11** 200?   **12** 99?   **13** 199?

211 take away 10 is 201.

201 is ten less than 211.

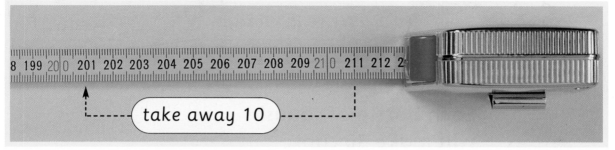

8 199 200 201 202 203 204 205 206 207 208 209 210 211 212 2

take away 10

What is ten less than:

14 134?　　**15** 243?　　**16** 215?　　**17** 107?

What is twenty less than:

18 99?　　**19** 100?　　**20** 200?　　**21** 300?

What is a hundred less than:

22 153?　　**23** 186?　　**24** 200?　　**25** 272?

Practise counting in jumps of ten, forwards and backwards.

Copy and complete.

26 51, 61, 71, 81, ____, ____, 111, 121, ____, ____.

27 294, 284, 274, 264, ____, ____, 234, 224, ____, ____.

28 157, 167, 177, 187, ____, ____, 217, 227, ____, ____.

Ask your teacher if you can play the '200 Trail' games.

Refreshments

We're in charge of refreshments!

SCHOOL FAIR this Saturday

Each bottle of squash will make 30 cups.

How many cups will:

1 3 bottles make?

2 6 bottles make?

Each box of tea bags will make 40 cups of tea.

How many cups will:

3 3 boxes make?

4 6 boxes make?

Each jar of coffee will make 35 cups.

How many cups will:

5 2 jars make?

6 4 jars make?

Each loaf of bread
will make 22 sandwiches.

How many sandwiches will:

7 3 loaves make?

8 5 loaves make?

9 10 loaves make?

10

We want to make
<u>about</u> 150 sandwiches.

How many loaves do we need?

Each pizza will cut
into 6 slices.

How many slices will:

11 4 pizzas make?

12 8 pizzas make?

13 10 pizzas make?

14

We want <u>about</u>
100 slices of pizza.

How many pizzas do we need?

Arithmetic within 300

Dice sums

This dice has ten numbers on it.

I threw the dice 4 times. I got these numbers:

I made up these sums.

Copy and complete.

1   50
       + 74
      _____

2   70
       + 54
      _____

3   75
       + 40
      _____

4   40
         7
      + 5
      _____

5 5 + 70 + 4

6 7 + 5 + 4 + 0

I threw the dice 3 times. I added the numbers in my head.

 9 add 4 is 13 …

add 3 is 16.

Add these in your head.

7

8

9

10

≈ Dice race ≈

A game for 2 people.

You need: two 0 to 9 dice and two calculators.

Ready, steady, GO!

Throw your dice ten times.

Write the numbers.

Add them in your head.

Write the total.

5 + 1 + 3 + 0 + 3 + 4 + 6 + 9 + 1 + 7 = **39**

Check

Check each other's totals with a calculator.

Have five turns ... or more!

School fair

We sold cakes at the fair.

Here is the money we made. £37

1 How much money?

2 How much money?

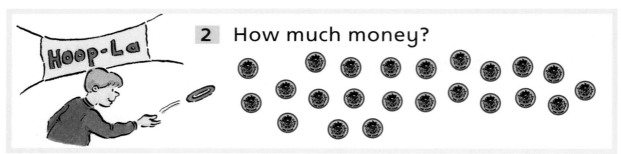

3 How much money?

4

Was it _easier_ to count the Hoop-La money or the Duck Game money?

Why?

5 How much money?

6 How much money?

7 How much money for toys, plants and books, altogether?

The raffle was 50p a ticket.

230 tickets were sold.

8 How much money?

The tombola was 10p a ticket.

400 tickets were sold.

9 How much money?

Ask your teacher if you can play the 'First to £500' game.

Tables superstars

I used tables stars in the Yellow books to practise my tables up to 6 × 6.

Superstars go up to 10 × 10.

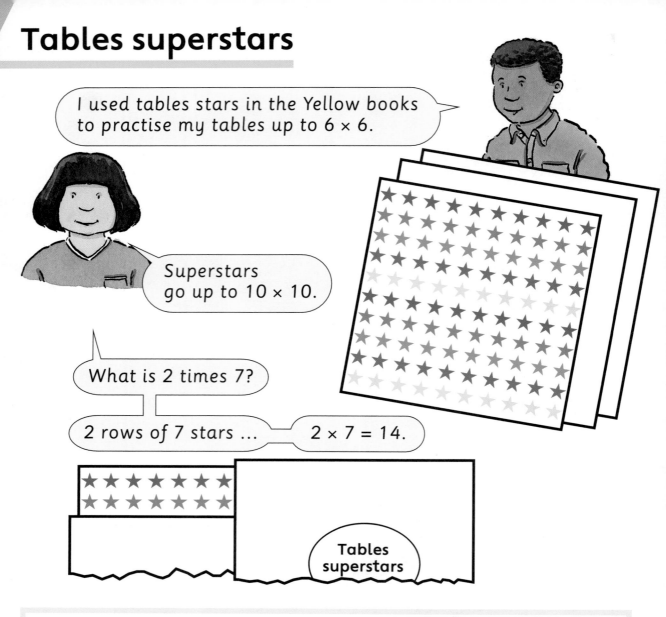

What is 2 times 7?

2 rows of 7 stars …

2 × 7 = 14.

Tables superstars

Make your own tables superstars from Copymaster P7.

Use your tables superstars.

1	2 × 8	**4**	3 × 8	**7**	6 × 7
2	10 × 9	**5**	4 × 5	**8**	0 × 8
3	7 × 3	**6**	9 × 2	**9**	10 × 10

≈ Superstars and dice ≈

A game for 2 people.

You need: a 0 to 9 dice, a 1 to 6 dice, and the tables superstars cards.

Which tables facts do you know?
Throw the dice.
Multiply the numbers.

5 times 7 is 35.

Your friend checks with the tables superstars like this.

Yes, 5 × 7 = 35.

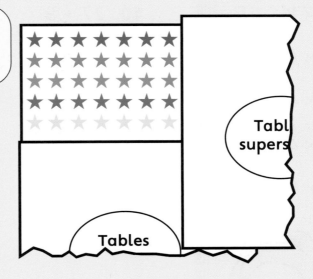

Tabl
supers

Tables

Have ten turns each ... or more!

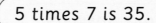

Tables facts within 100

▶ Copymasters P7 and P8

Speedy tables

Work with a friend.

Write question numbers 1 to 20.
Ask your friend to read the questions to you.

Write your answers
as quickly as you can.

1	5 × 10	**8**	90 ÷ 10	**15**	4 × 9
2	7 × 2	**9**	35 ÷ 5	**16**	7 ÷ 7
3	0 × 8	**10**	27 ÷ 9	**17**	3 × 8
4	3 × 6	**11**	21 ÷ 3	**18**	42 ÷ 7
5	8 × 6	**12**	18 ÷ 2	**19**	1 × 6
6	4 × 4	**13**	28 ÷ 4	**20**	24 ÷ 4
7	6 × 5	**14**	36 ÷ 6		

 ✓ or ✗

Now <u>you</u> read the questions
to your friend.

Mental recall of tables facts; all of the 0, 1, 2, 3, 4, 5, 6 and 10 times tables

Copy and complete.

21 ☐ × 5 = 0   **24** 3 × ☐ = 12

22 7 × ☐ = 35   **25** 6 × ☐ = 60

23 ☐ × 9 = 45   **26** ☐ × 5 = 25

How many 6s make 24?

6s into 24 ... that's 4.

27 18 ÷ 6   **30** 48 ÷ 6   **33** 60 ÷ 6

28 42 ÷ 6   **31** 6 ÷ 6   **34** 12 ÷ 6

29 30 ÷ 6   **32** 24 ÷ 6   **35** 54 ÷ 6

Use a stopwatch or a sand timer.

Use Speedy tables G made from Copymaster P9.

Speedy tables G
1 2 3 minute test
Name _____
Date _____

3 × 5 = ____   24 ÷ 8 = ____   0 × 4 = ____
6 × 7 = ____   21 ÷ 3 = ____   24 ÷ 4 = ____
4 × 8 = ____   40 ÷ 5 = ____   5 × 6 = ____
10 × 10 = ____   18 ÷ 6 = ____   28 ÷ 7 = ____
9 × 4 = ____   48 ÷ 8 = ____   6 × 9 = ____
7 × 5 = ____   16 ÷ 2 = ____   45 ÷ 9 = ____
3 × 9 = ____   25 ÷ 5 = ____   Score: ____

Can you get 20 questions right in 3 minutes?

Talk to your teacher about what to do next.

Check your answers. ✓ or ✗
Count how many you got right.

Fractions

We're sharing this between 5.

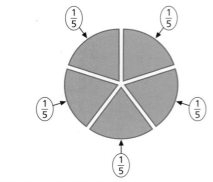

If you cut something into 5 pieces <u>the same size</u>, each piece is called a <u>fifth</u>.

Is each piece $\frac{1}{5}$? Write <u>Yes</u> or <u>No</u>.

1
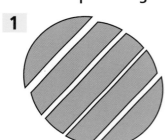

2

3

What fractions are these cakes cut into?

4
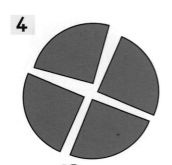

5

6
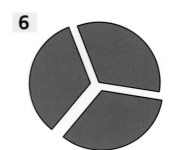

7

I cut my cake into sixths. How many pieces do you think there were?

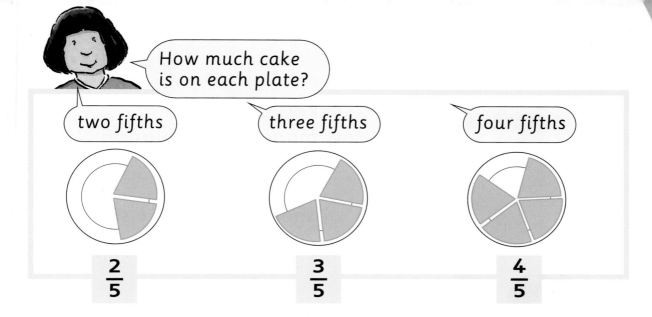

How much cake is on each plate?

two fifths — $\dfrac{2}{5}$

three fifths — $\dfrac{3}{5}$

four fifths — $\dfrac{4}{5}$

How much cake is on each plate?

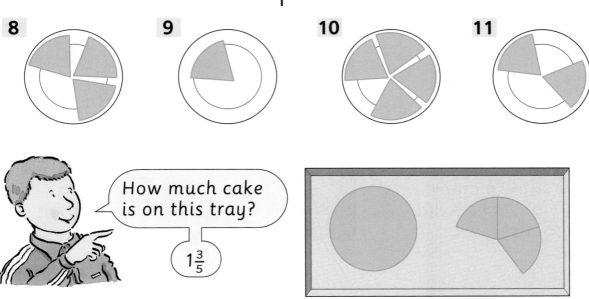

8

9

10

11

How much cake is on this tray?

$1\dfrac{3}{5}$

How much cake is on each tray?

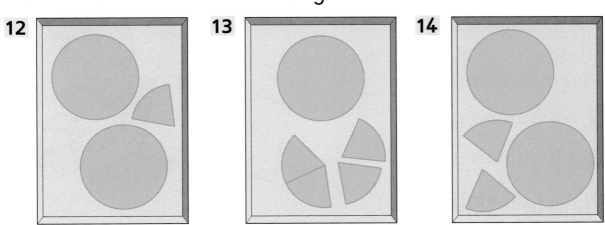

12

13

14

Stickers

6 stickers on a sheet.
How many guinea pigs on:

1 2 sheets?   **2** 3 sheets?

3 4 sheets?   **4** 6 sheets?

5 7 sheets?   **6** 9 sheets?

7 How many guinea pigs in my picture?

8 How many sheets of stickers did I use?

9 I want 30 guinea pig stickers. How many sheets do I need?

How many sheets for:

10 48 guinea pigs?

11 60 guinea pigs?

12 66 guinea pigs?

13 How many dogs on
one sheet of stickers?

How many dogs on:

14 2 sheets? **15** 3 sheets?

16 4 sheets? **17** 5 sheets?

$$\begin{array}{r} 12 \\ \times\ 6 \\ \hline 2 \\ \hline \end{array}$$

How many dogs on:

18 6 sheets? **19** 8 sheets?

20 9 sheets? **21** 10 sheets?

22  I had 84 dogs.
How many sheets
of stickers?

How many sheets for:

23 180 dogs? **24** 144 dogs?

More tables superstars

Use tables superstars cards made from Copymaster P7.

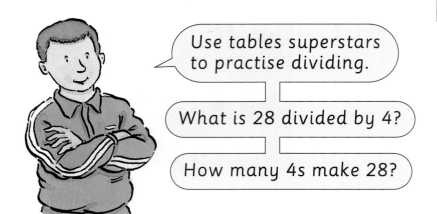

Use tables superstars to practise dividing.

What is 28 divided by 4?

How many 4s make 28?

6 lots of 4 is not enough.

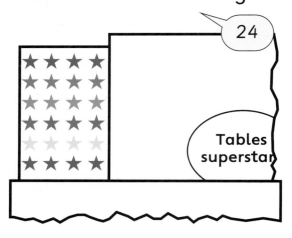

24

Tables superstar

It's 7 lots of 4.

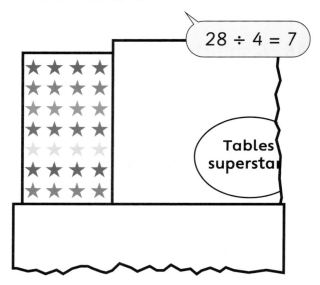

$28 \div 4 = 7$

Tables superstar

Use your tables superstars.

1 $36 \div 4$ **2** $18 \div 2$ **3** $24 \div 3$

4 $45 \div 9$ **5** $60 \div 10$ **6** $40 \div 5$

7 How many 8s make 16?

8 How many 9s make 81?

9 How many 5s make 40?

Sometimes you have stars left over.

What is 14 divided by 5?

How many 5s make 14?

2 lots of 5 is not enough.

3 lots of 5 is too many.

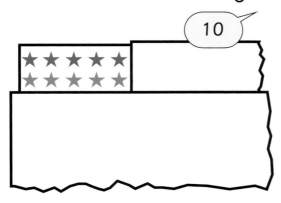

10

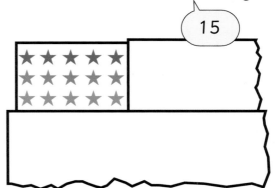

15

$14 \div 5 = 2 \text{ r.4}$

remainder

★ ★ ★ ★ ★
★ ★ ★ ★ ★
★ ★ ★ ★

$$\begin{array}{r} 2 \text{ r.4} \\ 5 \overline{)\,14} \end{array}$$

Copy and complete.

10
$$5 \overline{)\,12}$$

11
$$5 \overline{)\,20}$$

12
$$2 \overline{)\,16}$$

13
$$2 \overline{)\,17}$$

14
$$3 \overline{)\,12}$$

15
$$3 \overline{)\,14}$$

16
$$5 \overline{)\,42}$$

17
$$10 \overline{)\,34}$$

18
$$10 \overline{)\,68}$$

Nine times table

0, 9, 18, 27, 36, 45, ...

These numbers are called <u>multiples of nine</u>.

You can get <u>multiples of nine</u> by adding nines, or by <u>multiplying</u> by nine.

1 How many cubes?

2 How many cubes?

3 How many cubes?

It's 20 take away 2.

4 9 + 9

5 $2 \times 9 =$

6 How many cubes?

7 9 + 9 + 9 + 9

8 $4 \times 9 =$

9 How many cubes?

It's 60 take away 6.

10 9 + 9 + 9 + 9 + 9 + 9   **11** 6 × 9 =

63 cubes!

12 How many nines make 63?

13 6 3 ÷ 9 =

14 9 + 9 + 9 + 9 + 9 + 9 + 9 + 9   **15** 8 × 9 =

16 Copy and complete.

Nine times table

0 × 9 =	4 × 9 =	8 × 9 =
1 × 9 =	5 × 9 =	9 × 9 =
2 × 9 =	6 × 9 =	10 × 9 =
3 × 9 =	7 × 9 =	

Multiplying

42 times 3

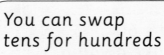

You can swap tens for hundreds if you want to.

$$\begin{array}{r} 42 \\ \times\ 3 \\ \hline 126 \\ \hline \end{array}$$

Make 3 lots of each number with tens and ones.

Copy and complete.

Forty-one

1
$$\begin{array}{r} 41 \\ \times\ 3 \\ \hline \end{array}$$

Sixty-two

2
$$\begin{array}{r} 62 \\ \times\ 3 \\ \hline \end{array}$$

Fifty-three

3
$$\begin{array}{r} 53 \\ \times\ 3 \\ \hline \end{array}$$

Eighty

4
$$\begin{array}{r} 80 \\ \times\ 3 \\ \hline \end{array}$$

34 times 3

34 times 5

That's
30 times 5
and
4 times 5.

150 + 20

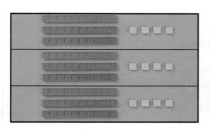

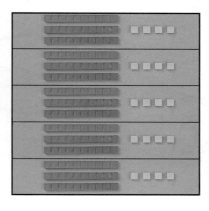

Copy and complete.

5
```
  34
×  3
─────
```

6
```
  34
×  4
─────
```

7
```
  34
×  5
─────
```

Use tens and ones.

8
```
  49
×  2
─────
```

9
```
  53
×  2
─────
```

10
```
  65
×  2
─────
```

11
```
  48
×  3
─────
```

12
```
  39
×  3
─────
```

13
```
  55
×  4
─────
```

14
```
  72
×  3
─────
```

15
```
  19
×  5
─────
```

Crazy golf

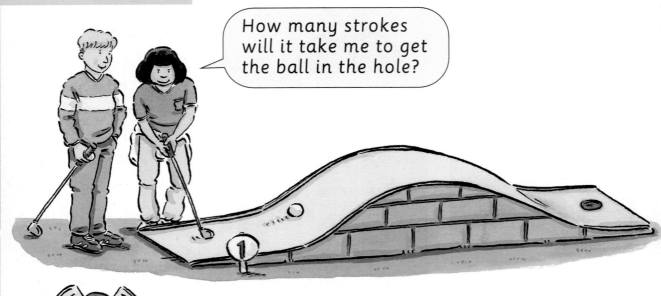

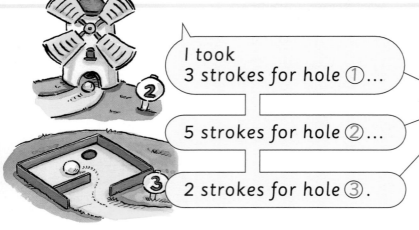

CRAZY GOLF			
Name: Wei			
HOLE	STROKES	HOLE	STROKES
1	3	7	4
2	5	8	1
3	2	9	3
4	7	10	2
5	3	11	5
6	2	12	4
		TOTAL	

1 How many strokes for hole ⑥?

2 How many strokes for hole ⑩?

3 Which hole was a 'hole in one'?

4 Which hole was the hardest for Wei?

5 What was her total score?

Add these in your head.
Check by adding in a different order.

6

CRAZY GOLF
Name: Owen

HOLE	STROKES	HOLE	STROKES
1	3	7	12
2	3	8	3
3	3	9	2
4	2	10	3
5	5	11	1
6	4	12	5
		TOTAL	

7

CRAZY GOLF
Name: Tanya

HOLE	STROKES	HOLE	STROKES
1	2	7	7
2	4	8	2
3	2	9	3
4	2	10	4
5	3	11	5
6	5	12	2
		TOTAL	

8

CRAZY GOLF
Name: Raphael

HOLE	STROKES	HOLE	STROKES
1	6	7	4
2	3	8	2
3	2	9	1
4	2	10	3
5	3	11	4
6	2	12	3
		TOTAL	

9 Who was the winner: Wei, Owen, Tanya or Raphael?

10

What did David score on hole ⑦?

CRAZY GOLF
Name: David

HOLE	STROKES	HOLE	STROKES
1	2	7	
2	3	8	3
3	4	9	4
4	3	10	2
5	2	11	2
6	1	12	3
		TOTAL	48

CRAZY GOLF
Name: Tom

HOLE	STROKES	HOLE	STROKES
1	4	7	3
2	3	8	2
3	3	9	4
4	3	10	5
5	3	11	3
6		12	3
		TOTAL	40

11 What did Tom score on hole ⑥?

Dividing by 2 or 3

We've got £10 notes and £1 coins.

We'll share them between the two of us.

1

£46 divided by 2

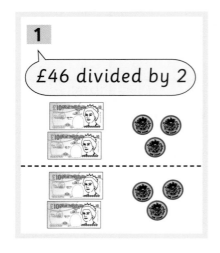

2

£56 divided by 2

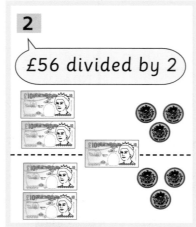

3

£66 divided by 2

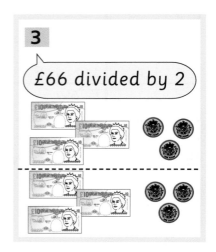

Copy and complete.

4 $2\overline{)46}$

5 $2\overline{)56}$

6 $2\overline{)66}$

7 $4\ 6 \div 2 =$

8 $5\ 6 \div 2 =$

9 $6\ 6 \div 2 =$

10 £86 divided by 2

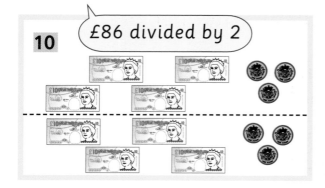

11 £96 divided by 2

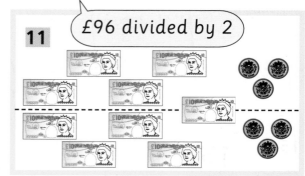

There are lots of ways of doing a division.

84 ÷ 3

We use tens and ones.

Two tens each ...   and two tens left, to swap for ones.

60 in tens,
so
≋20 each≋

24 in ones,
so
≋8 each≋

$$84 ÷ 3 = 28$$

This is how I divide on paper.

3s into 8 goes 2 ...   and 2 left over.   3s into 24 goes 8.

$$3 \overline{) 8\ 4}$$ (2)

$$3 \overline{) 8\,^2 4}$$ (2)

$$3 \overline{) 8\,^2 4}$$ (2 8)

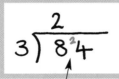

That's 2 tens left over.

Choose how to do these. Talk about what you did.

12 $3 \overline{) 93}$

13 $3 \overline{) 72}$

14 $3 \overline{) 78}$

15 $3 \overline{) 87}$

16 $3 \overline{) 69}$

17 $3 \overline{) 81}$

≈ What's my number? ≈

A game for 2, 3 or 4 people.

You need: cards numbered 1 to 100.

Shuffle the cards.
Put them in a pile, face down.

Take the top card.
Don't tell anyone the number.

Try to guess your friend's number.
Ask questions to help you.

For example,
  Is it bigger than 50?
    Is it an odd number?
      Is it in the ten times table?

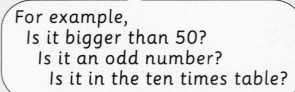

I can only answer
<u>Yes</u> or <u>No</u>.

Count how many questions
your friends ask before
they guess your number.

Have three turns each ... or more!

Part 2
Contents

Counting and place value
Addition and subtraction
Multiplication and division
Mixed problems

Eight hundred

How many is this?

245

---

1 How many is this?

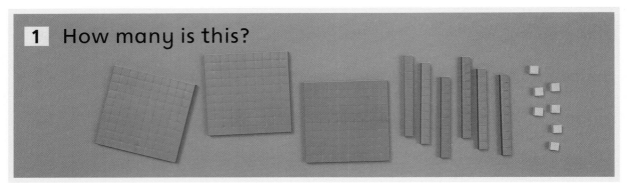

---

2 How many?

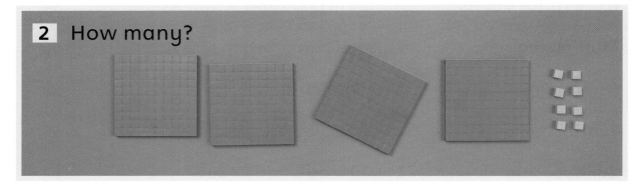

---

3 How many?

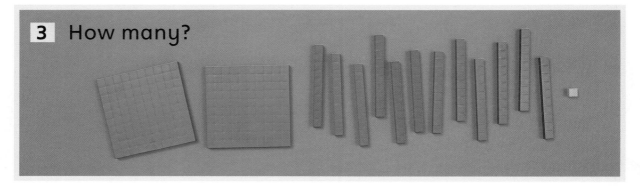

Work with a friend.

Make each number
with hundreds, tens and ones.

I'll check
your number.

Yes, that's 264.

Then draw it.

264

4 351

5 400

6 199

7 276

8 505

9 550

10 743

11 101

12 616

13 660

14 333

15 72

16

Seven hundred and
twenty-two

17

One hundred
and nine

Making one hundred

How much change?

1		

2		

3		

4		

Use a number line ...

You've got to 35 ... How many more to get to 100?

Copy and complete.

5 $35 + \boxed{} = 100$

6 $72 + \boxed{} = 100$

7 $44 + \boxed{} = 100$

8 $29 + \boxed{} = 100$

100 runs is called a <u>century</u> in cricket.

I need 2 more runs for a century.

9 8

How many more runs for a <u>century</u>?

9

| | | 7 |

10

| | 2 | 2 |

11

| | 4 | 6 |

12

| | 6 | 1 |

13

| | 7 | 3 |

14

| | 8 | 9 |

Copy and complete.

15 30 + ☐ = 100

17 60 + ☐ = 100

16 34 + ☐ = 100

18 68 + ☐ = 100

Ask your teacher if you can play 'One hundred bingo'.

Bags and boxes

50 straws in a bag.

How many straws in:

1 2 bags? **2** 4 bags? **3** 5 bags?

4 6 bags? **5** 8 bags? **6** 9 bags?

7 I want 150 straws. How many bags do I need?

8 I want 500 straws. How many bags do I need?

Copy and complete.

9 5, 10, 15, 20, _____ , 30, _____ , 40, _____ , _____ .

10 50, 100, _____ , 200, _____ , 300, _____ , _____ , 450, _____ .

11 450, 400, _____ , _____ , 250, 200, _____ , _____ , 50.

12
$$\begin{array}{r} 50 \\ \times\ 3 \\ \hline \\ \hline \end{array}$$

13
$$\begin{array}{r} 50 \\ \times\ 7 \\ \hline \\ \hline \end{array}$$

14
$$\begin{array}{r} 50 \\ \times\ 2 \\ \hline \\ \hline \end{array}$$

15
$$\begin{array}{r} 50 \\ \times\ 5 \\ \hline \\ \hline \end{array}$$

50 chalks in a box.
How many chalks in:

16 7 boxes?   **17** 4 boxes?

18

I want 250 chalks.
How many boxes
do I need?

Each box has 5 blue chalks.

19 How many blue chalks in 20 boxes?

Each box has 25 white chalks.

20 How many white chalks in 10 boxes?

21 How many white chalks in 20 boxes?

5 stones in a box.
How many stones in:

22 7 boxes?

23 10 boxes?

24 20 boxes?

25 15 boxes?

26 35 boxes?

Arithmetic within 500

► Copymasters P34 and P35

Raffle tickets

This book of raffle tickets has all the numbers from 1 to 500.

186 186
187 187
188 188
189 189
190 190

Twice!

There are 5 different numbers on each page.

1 What do you notice about the last ticket on each page?

Copy these. Write the next five numbers.

2 5, 10, 15, 20, …

3 45, 50, 55, 60, …

4 105, 110, 115, 120, …

5 280, 285, 290, 295, …

Are these numbers multiples of 5? Write <u>Yes</u> or <u>No</u>.

6 135   **7** 136   **8** 299   **9** 300

I used tickets 1 to 100 to check my spelling.

36 *thirty-six*

Write these numbers as words.

10 **42**

11 **71**

12 **94**

13 **18**

14 **55**

15 **83**

I used tickets 1 to 100 to make a hundred square.

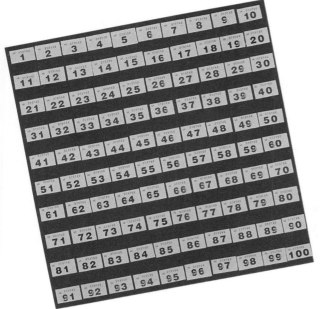

On this hundred square ...

16 How many tickets in each row?

17 How many tickets in each column?

18 How many numbers are multiples of ten?

19 How many numbers are even?

20 How many tickets have 7 on them?

Hundred squares

It is easy to add or take away in tens, using a hundred square.

27 + 10

52 – 10

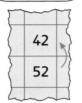

35 + 20

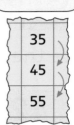

Do these using this hundred square:

1 32 + 10

2 49 – 10

3 26 + 20

4 48 + 30

5 85 – 50

1	2	3	4	5	6	7	8	9	10
11	12	13	14	15	16	17	18	19	20
21	22	23	24	25	26	27	28	29	30
31	32	33	34	35	36	37	38	39	40
41	42	43	44	45	46	47	48	49	50
51	52	53	54	55	56	57	58	59	60
61	62	63	64	65	66	67	68	69	70
71	72	73	74	75	76	77	78	79	80
81	82	83	84	85	86	87	88	89	90
91	92	93	94	95	96	97	98	99	100

You can add or take away smaller numbers, too.

25 + 3

25	26	27	28

17 – 1

16	17

6 73 + 3

7 46 + 40

8 89 – 30

9 27 + 30

10 42 + 7

11 42 + 8

12 42 + 9

13 12 + 80

14 65 – 50

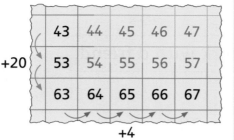

Instead of adding 43 + 24,
I can add 43 + 20 + 4, or 43 + 4 + 20.

	43	44	45	46	47
+20	53	54	55	56	57
	63	64	65	66	67

+4

+4

	43	44	45	46	47	
	53	54	55	56	57	+20
	63	64	65	66	67	

Use the 1 to 100 square.

15 26 + 30 + 3

16 26 + 3 + 30

17 26 + 33

18 32 + 50 + 4

19 32 + 4 + 50

20 32 + 54

21 37 + 52

22 29 + 12

23 65 + 33

This hundred square
goes from 251 to 350.

251	252	253	254	255	256	257	258	259	260
261	262	263	264	265	266	267	268	269	270
271	272	273	274	275	276	277	278	279	280
281	282	283	284	285	286	287	288	289	290
291	292	293	294	295	296	297	298	299	300
301	302	303	304	305	306	307	308	309	310
311	312	313	314	315	316	317	318	319	320
321	322	323	324	325	326	327	328	329	330
331	332	333	334	335	336	337	338	339	340
341	342	343	344	345	346	347	348	349	350

24 263 + 20

25 294 + 30

26 326 – 50

27 309 – 20

28 298 + 5

29 344 – 50

30 283 + 45

31 339 – 60

32 311 + 26

33 280 + 38

34 277 + 54

35 350 – 51

Speedy tables

Work with a friend.

Write question numbers 1 to 20.
Ask your friend to read the questions to you.

Write your answers
as quickly as you can.

1	3 × 6
2	9 × 4
3	1 × 1
4	7 × 6
5	5 × 9
6	2 × 9
7	9 × 9

8	16 ÷ 2
9	54 ÷ 9
10	24 ÷ 3
11	30 ÷ 6
12	20 ÷ 5
13	50 ÷ 10
14	35 ÷ 5

15	7 × 9
16	32 ÷ 4
17	6 × 6
18	27 ÷ 9
19	8 × 6
20	28 ÷ 4

 ✓ or ✗

Now you read the questions
to your friend.

Mental recall of tables facts; all of the 0, 1, 2, 3, 4, 5, 6, 9 and 10 times tables

Copy and complete.

21 $3 \times \boxed{} = 18$ **24** $6 \times \boxed{} = 42$

22 $\boxed{} \times 6 = 0$ **25** $\boxed{} \times 6 = 54$

23 $6 \times 6 = \boxed{}$ **26** $\boxed{} \times 6 = 24$

How many 9s make 27?

9s into 27? That's 3.

27 $9 \div 9$ **30** $45 \div 9$ **33** $27 \div 9$

28 $36 \div 9$ **31** $18 \div 9$ **34** $72 \div 9$

29 $54 \div 9$ **32** $63 \div 9$ **35** $90 \div 9$

Use a stopwatch or a sand timer.

Use Speedy tables **I** made from Copymaster P40.

Speedy tables [I]
1 2 3 minute test
Name _____
Date _____

$4 \times 6 =$ ___ $27 \div 3 =$ ___ $8 \times 9 =$ ___
$5 \times 3 =$ ___ $9 \div 9 =$ ___ $16 \div 4 =$ ___
$2 \times 7 =$ ___ $40 \div 5 =$ ___ $0 \times 8 =$ ___
$9 \times 9 =$ ___ $21 \div 7 =$ ___ $42 \div 6 =$ ___
$8 \times 6 =$ ___ $63 \div 9 =$ ___ $10 \times 4 =$ ___
$5 \times 5 =$ ___ $48 \div 6 =$ ___ $54 \div 6 =$ ___
$7 \times 4 =$ ___ $70 \div 10 =$ ___   Score:

Can you get 20 questions right in 3 minutes?

Talk to your teacher about what to do next.

Check your answers. ✓ or ✗
Count how many you got right.

Hoops

We score 7 points for each hoop over a post.

What did we score?

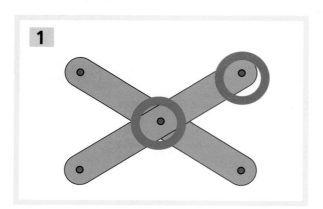

1

2

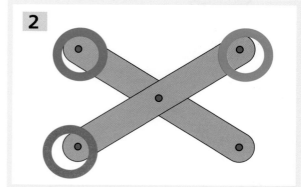

3

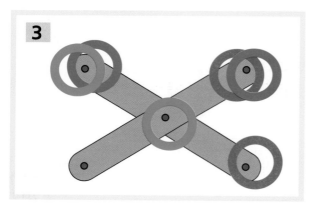

4

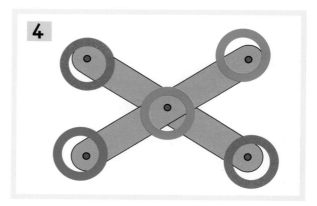

5

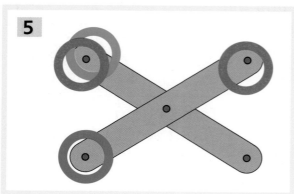

6

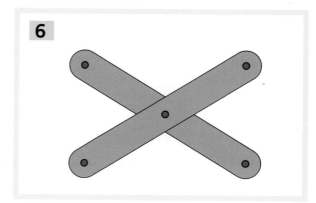

7 I scored 14.

How many hoops?

8 I scored 35.

How many hoops?

9 I scored 28.

How many hoops?

10 I scored 7.

How many hoops?

11 I scored 42.

How many hoops?

12 I scored 49.

How many hoops?

Copy and complete.

13 3 × 7

14 5 × 7

15 2 × 7

16 4 × 7

17 14 ÷ 7

18 35 ÷ 7

19 21 ÷ 7

20 42 ÷ 7

Check. ✓ or ✗

Ask your teacher if you can play the 'Forty-nine' game.

What's the difference?

I'm 24.

I'm 13 years old. What's the difference in our ages?

24 − 13 = 11

11 years difference.

What is the difference in these ages?

1

I'm 35.

I'm 38.

2

I'm 15.

I'm 43.

3

I'm 3.

I'm 102.

4

I'm 75.

I'm 56.

5

I'm 9.

There's 22 years difference in our ages.

How old am I?

When he was a kitten, Ginger weighed 1 kilogram.

Now he weighs 7 kg.

6 What is the difference in Ginger's weight?

Posh Shampoo 98 ml £2·50

Cheap Shampoo = 500 ml £2·60

7 What is the difference between the amounts of shampoo in these bottles?

8 What is the difference in their prices?

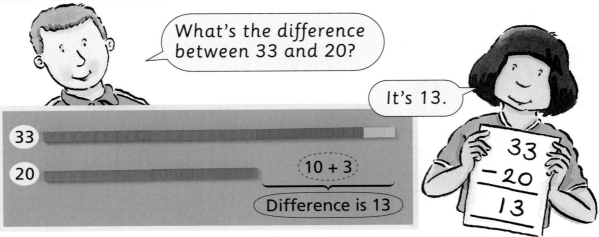

What's the difference between 33 and 20?

It's 13.

33
20
10 + 3
Difference is 13

33
−20
13

What is the difference between:

9 49 and 86?

10 150 and 450?

11 120 and 35?

12 50 and 500?

13 300 and 30?

 ✓ or ✗

More raffle tickets

I want to practise adding.

Pick the tickets ... then add the numbers.

Write each sum and add up.

1 141 227

2 231 59

3 90 173

4 65 242

5 104 53 216

Put the biggest number first.
Then practise taking away!

6 216 153

7 171 120

8 250 149

9 6 103

≈ Raffle sums ≈

A game for 2 people.

You need: raffle tickets numbered 1 to 200, a tub and a calculator.

Put the tickets in the tub.

Take two tickets from the tub.

Add the numbers in your head or on paper.

207.

Your friend checks on the calculator.

Yes, 181 + 26 = 207.

Have ten turns each ... or more!

Addition and subtraction within 500   ▶ Copymasters P45 and P46   **55**

Multiplying

Remember, swap if you want to.

and

162 times 2

$$\begin{array}{r} 1\ 6\ 2 \\ \times\qquad 2 \\ \hline 3\ 2\ 4 \\ \hline \end{array}$$

Make 2 lots of each number with hundreds, tens and ones.

Copy and complete.

One hundred and forty-seven

1
$$\begin{array}{r} 147 \\ \times\quad 2 \\ \hline \end{array}$$

Two hundred and sixteen

2
$$\begin{array}{r} 216 \\ \times\quad 2 \\ \hline \end{array}$$

One hundred and eighty-five

3
$$\begin{array}{r} 185 \\ \times\quad 2 \\ \hline \end{array}$$

One hundred and seventy-nine

4
$$\begin{array}{r} 179 \\ \times\quad 2 \\ \hline \end{array}$$

125 times 2

125 times 4

That's double and double again!

Double 125 is 250.
Double 250 is 500.

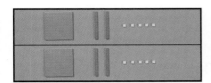

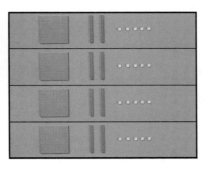

Copy and complete.

5
```
  125
×   2
─────
```

6
```
  125
×   3
─────
```

7
```
  125
×   4
─────
```

Use hundreds, tens and ones.
Then multiply on paper.

8
```
  246
×   2
─────
```

9
```
  171
×   2
─────
```

10
```
  140
×   3
─────
```

11
```
  106
×   3
─────
```

12
```
  105
×   4
─────
```

13
```
   93
×   4
─────
```

14
```
   70
×   5
─────
```

15
```
   74
×   5
─────
```

Multiplication within 500

Seven times table

0, 7, 14, 21, 28, 35, ...

These numbers are called <u>multiples of seven</u>.

You can get <u>multiples of seven</u> by adding sevens, or by <u>multiplying</u> by seven.

Two boxes

1 How many cakes?

2 7 + 7

3 **2** × **7** =

Three boxes

4 How many cakes?

5 7 + 7 + 7

6 **3** × **7** =

Four boxes

7 How many cakes?

8 7 + 7 + 7 + 7

9 **4** × **7** =

Six boxes

10 How many cakes?

11 7 + 7 + 7 + 7 + 7 + 7

12 **6** × **7** =

Seven boxes

13 How many cakes?

14 7 × 7 =

Fifty-six cakes

15 How many boxes?

Copy and complete.

16 56 = 7 × ☐   **5** **6** **7** **8**

Nine boxes

17 How many cakes?

18 9 × 7 =

19 Copy and complete.

Seven times table

0 × 7 =	4 × 7 =	8 × 7 =
1 × 7 =	5 × 7 =	9 × 7 =
2 × 7 =	6 × 7 =	10 × 7 =
3 × 7 =	7 × 7 =	

Tenths

We're sharing this between 10.

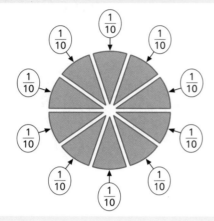

If you split something into 10 pieces <u>the same size</u>, each piece is called a <u>tenth</u>.

Is each cake cut into tenths? Write <u>Yes</u> or <u>No</u>.

1

2

3

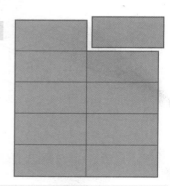

We'll have half a cake between us.

4 How many tenths is the same as $\frac{1}{2}$?

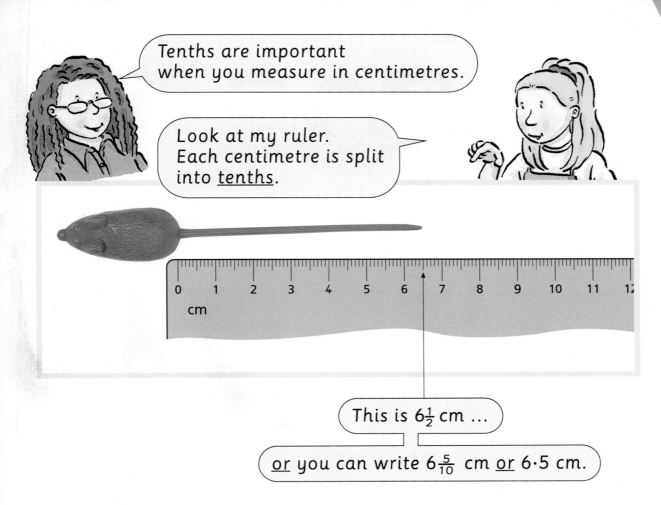

Tenths are important when you measure in centimetres.

Look at my ruler. Each centimetre is split into <u>tenths</u>.

This is $6\frac{1}{2}$ cm ...

<u>or</u> you can write $6\frac{5}{10}$ cm <u>or</u> 6·5 cm.

Measure each mouse's tail.
Write the length in the 3 different ways.

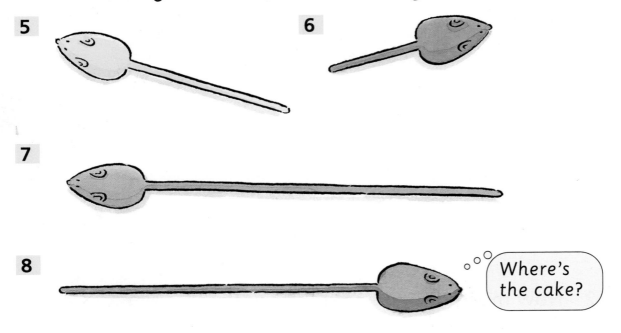

5

6

7

8

Where's the cake?

Car boot sale

> We sell things at a car boot sale once a month.

> Then we share the money between the three of us.

You need: £1 coins and £10 notes (Copymaster P29) and the chart made from Copymaster P53.

In January we made £129 altogether.

£129 ÷ 3 = £43

> £43 each

Check: 3 × £43 = £129 ✓

Use coins and notes.

February

> We made £141 altogether.

1 How much did each of us get?

March

> It was snowing! We only made £16 each.

2 How much did we make altogether?

I made a chart to help us share our money.

Ready reckoner for sharing money between 3 people

How much altogether?	How much each?	How much altogether?	How much each?
£30	£10	£1	33p
£60	£20	£2	66p
£90	£30	£3	£1
(£120)	£40	£6	£2
£150	£50	(£9)	£3
£180	£60	£12	£4

£1 doesn't divide exactly by 3. The closest you can get is 33p each.

In January we made £129 altogether. That's £120 and £9. So that's £40 and £3 each. £43 each.

Make your own ready reckoner with Copymaster P53.
Use your ready reckoner to do these:

April

3

£99 altogether. How much each?

May

4

£219 altogether. How much each?

June

5

£241 altogether. How much each?

Check with notes and coins.

What's missing?

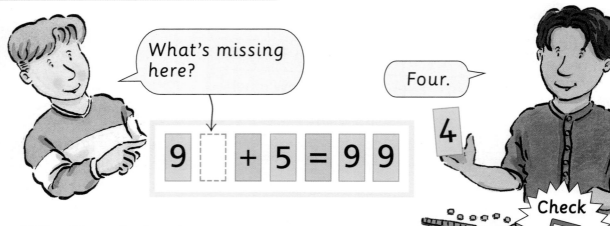

What's missing?

1 1 4 2 + ☐ 1 = 1 6 3

2 1 ☐ 4 − 6 2 = 1 2 2

3 5 7 + ☐ 7 = 1 1 4

4 1 3 4 + 1 ☐ 6 = 2 5 0

5 2 0 0 − 1 8 ☐ = 2 0

6 9 ☐ + 1 2 6 = 2 1 8

What's missing here?

Divided by.

$$4 \ 0 \ \boxed{} \ 2 = 2 \ 0$$

What's missing?

7
$$5 \ \boxed{} \ 5 = 2 \ 5$$

8
$$8 \ 1 \ \boxed{} \ 9 = 9$$

9
$$1 \ 2 \ 0 \ \boxed{} \ 2 = 1 \ 2 \ 2$$

10
$$1 \ 2 \ 0 \ \boxed{} \ 2 = 6 \ 0$$

11
$$7 \ \boxed{} \ 7 = 4 \ 9$$

12
$$4 \ 9 \ \boxed{} \ 7 = 7$$

13
$$6 \ 2 \ \boxed{} \ 3 = 1 \ 8 \ 6$$

14
$$3 \ 5 \ \boxed{} \ 1 \ 7 \ 0 = 2 \ 0 \ 5$$

15
$$2 \ 3 \ 2 \ \boxed{} \ 3 \ 8 = 1 \ 9 \ 4$$

≈ Double or half? ≈

A game for 2 or 3 people.

You need: cards numbered 1 to 100 and a calculator.

Shuffle the cards.
Put them in a pile, face down.

Take the top card.

If it is an odd number, double it.
If it is even, halve it.

Double 23 is 46.

Check
2 × 23 = 46 ✓

You were right, so you keep the card.

Now it is your friend's turn.

Have ten turns each ... or more!
Count how many cards you won.

Part 3
Contents

Counting and place value
Addition and subtraction
Multiplication and division
Mixed problems

One thousand

 130 add 180

 That's 310 altogether.

1 How many?

2 How many?

3 How many altogether?

4 How many?

5 How many?

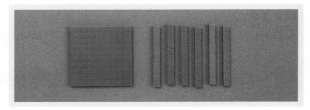

6 How many altogether?

7 How many?

8 How many?

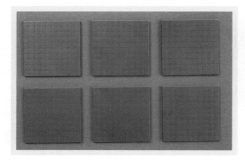

9 How many altogether?

Count in 10s.

Copy and complete.

10 80, 90, 100, 110, _____ , _____ , _____ , _____ .

11 350, 360, 370, 380, _____ , _____ , _____ , _____ .

12 990, 980, 970, 960, _____ , _____ , _____ , _____ .

Count in 50s.

13 50, 100, 150, 200, _____ , _____ , _____ , _____ .

14 450, 500, 550, 600, _____ , _____ , _____ , _____ .

15 950, 900, 850, 800, _____ , _____ , _____ , _____ .

Count in 20s.

16 100, 120, 140, 160, _____ , _____ , _____ , _____ .

17 280, 300, 320, 340, _____ , _____ , _____ , _____ .

18 440, 460, 480, 500, _____ , _____ , _____ , _____ .

19 740, 720, 700, 680, _____ , _____ , _____ , _____ .

20 980, 960, 940, 920, _____ , _____ , _____ , _____ .

Work with a friend.

Take turns in counting in 50s up to one thousand ... and down again!

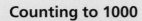

Rounding

This number line goes to 1000.

But there isn't room to write all the numbers.

What is this number?

867

Which number is each arrow pointing to?

1
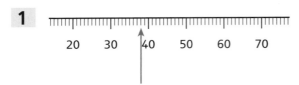

20   30   40   50   60   70

2

410   420   430   440   450   460

3
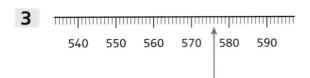

540   550   560   570   580   590

4

650   660   670   680   690   700

Make your own number line from Copymaster P62.

Work with a friend. Count backwards in tens from 1000.

1000, 990, 980, ...

Your friend can check on the number line.

I made 328 with hundreds, tens and ones ... and I found it on the number line.

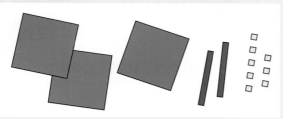

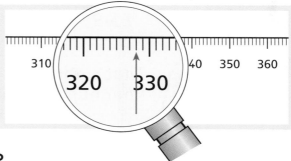

310 ... 320 330 ... 40 350 360

5 Is 328 closer to 320 or 330?

Sometimes you don't need an exact answer to a sum.

You can use numbers which are 'rounded'.

$$328 + 259$$

$$330 + 260 \over 590$$

Practise 'rounding to the nearest ten'.

Make each number ...   find it ...   round it to the nearest ten:

241      →   230 240 250 260   →   240

6 136

9 449

12 225

7 479

10 197

8 981

11 653

Talk to your teacher about this number!

Rough answers

I've spent £1·99 and £3·95.

That's about £2 and £4, to the nearest pound. Roughly £6 altogether.

Work out rough answers.

1

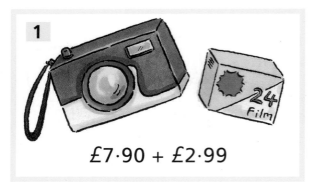

£7·90 + £2·99

2

£2·95 + £1·95

3

£3·85 + £3·20

4

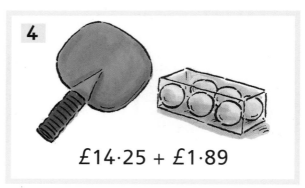

£14·25 + £1·89

5

£1·95 + £3·75

6

£2·25 + £1·90

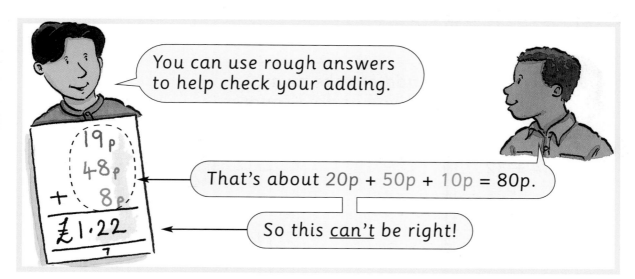

You can use rough answers to help check your adding.

That's about 20p + 50p + 10p = 80p.

So this <u>can't</u> be right!

	19p
	48p
+	8p
£1·22	

I used my calculator to do these.

Check with rough answers.

Do my answers seem reasonable? Write <u>Yes</u> or <u>No</u>.

7
$$£1·20$$
$$+ \quad 99p$$
$$\overline{£100·20}$$

8
$$54p$$
$$+37p$$
$$\overline{91p}$$

9
$$£4·99$$
$$+ £7·99$$
$$\overline{£12·08}$$

10 £1·05 + £23·99 + £6·95 + £2·99 = £28·03

11 £3·95 + £3·95 = £7·90   **12** 99p + 99p + 99p = £2·97

Ask your teacher if you can play the 'Rough total' game.

Hundred squares

I can add or take away, using two hundred squares.

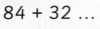

84 + 32 ...

that's 84 + 30 + 2

= 116

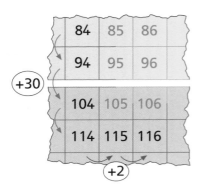

1	2	3	4	5	6	7	8	9	10
11	12	13	14	15	16	17	18	19	20
21	22	23	24	25	26	27	28	29	30
31	32	33	34	35	36	37	38	39	40
41	42	43	44	45	46	47	48	49	50
51	52	53	54	55	56	57	58	59	60
61	62	63	64	65	66	67	68	69	70
71	72	73	74	75	76	77	78	79	80
81	82	83	84	85	86	87	88	89	90
91	92	93	94	95	96	97	98	99	100

101	102	103	104	105	106	107	108	109	110
111	112	113	114	115	116	117	118	119	120
121	122	123	124	125	126	127	128	129	130
131	132	133	134	135	136	137	138	139	140
141	142	143	144	145	146	147	148	149	150
151	152	153	154	155	156	157	158	159	160
161	162	163	164	165	166	167	168	169	170
171	172	173	174	175	176	177	178	179	180
181	182	183	184	185	186	187	188	189	190
191	192	193	194	195	196	197	198	199	200

Use the hundred squares.

1 56 + 51

2 116 − 20

3 139 − 60

4 139 − 64

5 75 + 30

6 75 + 34

7 75 + 38

8 163 − 80

9 24 + 100

10 24 + 120

11 142 − 70

12 120 − 59

Now use <u>three</u> hundred squares.

13 314 + 50

14 362 + 57

15 328 − 70

16 446 − 90

17 394 + 87

18 353 + 67

19 288 − 51

20 404 − 21

What's missing?

Copy and complete.

21 300 + ☐ = 500

22 375 + ☐ = 500

23 382 + ☐ = 500

24 433 + ☐ = 500

201	202	203	204	205	206	207	208	209	210
211	212	213	214	215	216	217	218	219	220
221	222	223	224	225	226	227	228	229	230
231	232	233	234	235	236	237	238	239	240
241	242	243	244	245	246	247	248	249	250
251	252	253	254	255	256	257	258	259	260
261	262	263	264	265	266	267	268	269	270
271	272	273	274	275	276	277	278	279	280
281	282	283	284	285	286	287	288	289	290
291	292	293	294	295	296	297	298	299	300

301	302	303	304	305	306	307	308	309	310
311	312	313	314	315	316	317	318	319	320
321	322	323	324	325	326	327	328	329	330
331	332	333	334	335	336	337	338	339	340
341	342	343	344	345	346	347	348	349	350
351	352	353	354	355	356	357	358	359	360
361	362	363	364	365	366	367	368	369	370
371	372	373	374	375	376	377	378	379	380
381	382	383	384	385	386	387	388	389	390
391	392	393	394	395	396	397	398	399	400

401	402	403	404	405	406	407	408	409	410
411	412	413	414	415	416	417	418	419	420
421	422	423	424	425	426	427	428	429	430
431	432	433	434	435	436	437	438	439	440
441	442	443	444	445	446	447	448	449	450
451	452	453	454	455	456	457	458	459	460
461	462	463	464	465	466	467	468	469	470
471	472	473	474	475	476	477	478	479	480
481	482	483	484	485	486	487	488	489	490
491	492	493	494	495	496	497	498	499	500

Corner cards

You need a set of corner cards made from Copymaster P89.

Corner cards show hundreds, tens and ones.

How do you make three hundred and forty-one?

Collect the cards ... and put them together.

| 3 0 0 | 4 0 | 1 | → | 3 4 1 |

Make each number with corner cards.
Then write it down.

1 Four hundred and seventy-two

2 Five hundred and eighty-six

3 Two hundred and five

4 Nine hundred and twenty-four

5 Seven hundred and thirteen

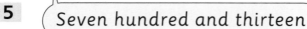

6 One hundred and sixty

7 Eight hundred and fifty-nine

Use corner cards to make up sums ...

choose a hundreds, a tens, and a ones card to make each number.

2 3 7

Add these:

237
+ 345

8

2 3 7

3 4 5

9

1 9 3

4 6 1

10

2 2 8

3 7 6

11

4 5 2

1 8 4

12

3 4 9

2 8 5

13

4 3 7

3 5 5

14

4 9 4

2 6 6

Ask your teacher if you can play the '999' game.

Boxes and bones

15 giant paperclips in a box.

How many paperclips in:

1 2 boxes?

2 3 boxes?

3 4 boxes?

4 8 boxes?

5 9 boxes?

6 I want 75 paperclips.
How many boxes do I need?

7 I want 150 paperclips.
How many boxes do I need?

Copy and complete.

8
$$\begin{array}{r} 15 \\ \times\ 6 \\ \hline \\ \hline \end{array}$$

9
$$\begin{array}{r} 15 \\ \times\ 7 \\ \hline \\ \hline \end{array}$$

10
$$\begin{array}{r} 15 \\ \times\ 5 \\ \hline \\ \hline \end{array}$$

11
$$\begin{array}{r} 15 \\ \times\ 10 \\ \hline \\ \hline \end{array}$$

60 push pins in a box.
How many pins in:

12 3 boxes? **13** 4 boxes?

14 6 boxes? **15** 10 boxes?

16

> I want 300 pins.
> How many boxes do I need?

> I need 10 paper fasteners
> to make a skeleton.

17

> How many skeletons could
> I make with one box
> of paper fasteners?

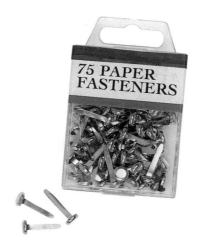

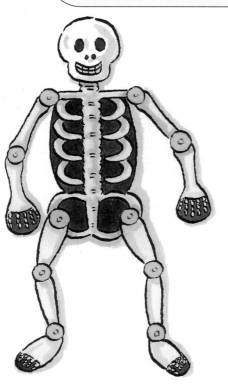

18 How many paper fasteners
in 2 boxes?

19 How many skeletons with
2 boxes of paper fasteners?

20 How many paper fasteners
in 10 boxes?

Stamps

I collect stamps.

1 Ten stamps on every page.
14 pages in the book.

How many stamps altogether?

I have 34 stamps of birds and 77 stamps of other animals.

2 How many stamps altogether?

I had 98 stamps, to start with.

My friend gave me this stamp.
I gave her 3 animal stamps.

3 How many stamps have I got now?

50 stamps in each pack.
How many stamps in:

4 3 packs?

5 5 packs?

6 10 packs?

How much would it cost for:

7 3 packs?   **8** 5 packs?   **9** 10 packs?

These stamps cost about 4p each.

No, they are about 3p each.

10 How much do <u>you</u> think they are?

You need <u>new</u> stamps to post letters!

10 stamps in a book.
How many stamps in:

11 8 books?   **12** 20 books?

Ask your teacher if you can play the 'Eights' game.

More corner cards

You need a set of corner cards made from Copymaster P89.

Corner cards help you think about adding.

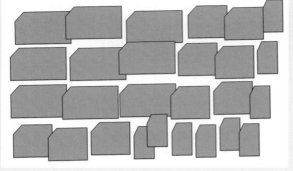

We chose these cards:

2 0   8 0   3   7

1 Add these:

2 3   8 7

2 Add these:

8 3   2 7

I chose these:

7 0   6 0   8   5

3 6 8 + 7 5

4 7 8 + 6 5

5 7 0 + 8 + 6 0 + 5

6 6 0 + 8 + 7 0 + 5

7 6 0 + 7 0 + 8 + 5

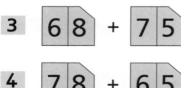

You can add in any order.
Choose your favourite way!

$$381 + 270$$

I'll do
300 + 200 + 80 + 70 + 1.

$$300 \quad 200 \quad 80 \quad 70 \quad 1$$

I'll do
70 + 80 + 300 + 200 + 1.

$$70 \quad 80 \quad 300 \quad 200 \quad 1$$

Make each number with corner cards.
Put them in your favourite order and add them.

8 $802 + 140$

9 $723 + 97$

10 $464 + 143$

Write down
what you do.

800 + 100 + 40 + 2

11 $58 + 92$

12 $230 + 670$

13 $66 + 904$

14 $515 + 380$

Eight times table

0, 8, 16, 24, 32, 40, ...

These numbers are called <u>multiples of eight</u>.

You can get <u>multiples of eight</u> by adding eights, or by <u>multiplying</u> by eight.

Two packs

1 How many cups?

2 8 + 8

3 $2 \times 8 =$

Three packs

4 How many cups?

5 8 + 8 + 8

6 $3 \times 8 =$

Five packs

7 How many cups?

8 8 + 8 + 8 + 8 + 8

9 $5 \times 8 =$

Six packs

10 How many cups?

11 8 + 8 + 8 + 8 + 8 + 8

12 $6 \times 8 =$

Seven packs

13 How many cups?

14 $7 \times 8 =$

Sixty-four cups

15 How many packs?

16 $64 \div 8 =$

Nine packs

17 How many cups?

18 $9 \times 8 =$

19 Copy and complete.

Eight times table

$0 \times 8 =$	$4 \times 8 =$	$8 \times 8 =$
$1 \times 8 =$	$5 \times 8 =$	$9 \times 8 =$
$2 \times 8 =$	$6 \times 8 =$	$10 \times 8 =$
$3 \times 8 =$	$7 \times 8 =$	

Add or take away

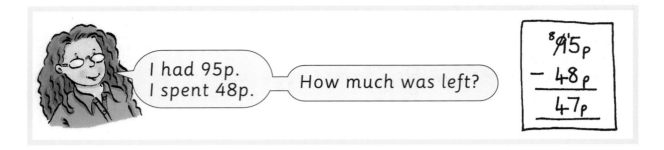

I had 95p. I spent 48p. How much was left?

$$\begin{array}{r} ^8\cancel{9}^{\,1}5p \\ -\ 48p \\ \hline 47p \end{array}$$

Write each sum and work it out.

1

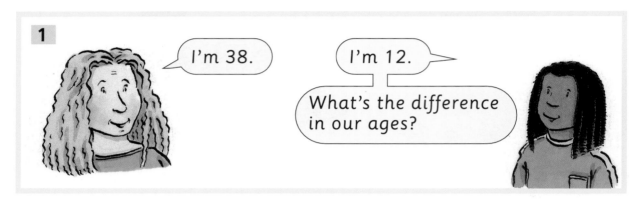

I'm 38.

I'm 12.

What's the difference in our ages?

2

24 comics here

118 comics here

How many altogether?

3

I had £5·70. I spent £3·49.

£7·00
£3·49

How much is left?

4

11 candles here

3 boxes of 25 candles here

How many altogether?

5

I'm 93 years old.
How old was I, 45 years ago?

6

I used to earn £4·50 for my
paper round.
Then I got an increase of £1·50.

How much do I earn now?

7

I've got £4·60,
plus £2·50
from my brother.

How much have I got?

8

I've got £10·00,
minus £1·25
for my bus fare.

How much have I got?

9

I've got £3·32.

I've got £1·70
more than you.

How much have I got?

Speedy tables

Work with a friend.

Write question numbers 1 to 20.
Ask your friend to read the questions to you.

Write your answers
as quickly as you can.

1	3 × 10	**8**	70 ÷ 10	**15**	9 × 7
2	5 × 7	**9**	28 ÷ 7	**16**	18 ÷ 6
3	9 × 4	**10**	32 ÷ 8	**17**	7 × 8
4	6 × 5	**11**	9 ÷ 9	**18**	16 ÷ 2
5	2 × 7	**12**	45 ÷ 5	**19**	7 × 0
6	8 × 8	**13**	24 ÷ 3	**20**	21 ÷ 3
7	7 × 6	**14**	20 ÷ 4		

 ✓ or ✗

Now <u>you</u> read the questions
to your friend.

Copy and complete.

21 $2 \times \boxed{} = 14$ **24** $7 \times \boxed{} = 49$

22 $\boxed{} \times 7 = 35$ **25** $\boxed{} \times 7 = 56$

23 $\boxed{} \times 7 = 0$ **26** $\boxed{} \times 7 = 28$

How many 8s make 32?

8s into 32 ... that's 4.

27 $80 \div 8$ **30** $24 \div 8$ **33** $72 \div 8$

28 $16 \div 8$ **31** $56 \div 8$ **34** $48 \div 8$

29 $40 \div 8$ **32** $32 \div 8$ **35** $64 \div 8$

 Use a stopwatch or a sand timer.

Use Speedy tables K made from Copymaster P80.

Can you get 20 questions right in 3 minutes?

Talk to your teacher about what to do next.

Check your answers. ✓ or ✗
Count how many you got right.

More multiplying

143 times 3

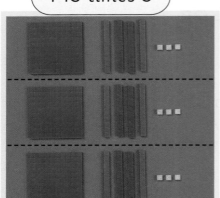

Remember, swap if you want to.

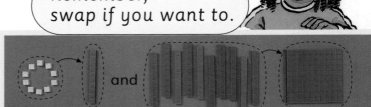

and

$$\begin{array}{r} 1\,4\,3 \\ \times \quad 3 \\ \hline 4\,2\,9 \\ \end{array}$$

Make 3 lots of each number with hundreds, tens and ones.

Copy and complete.

One hundred and thirty-six

1
$$\begin{array}{r} 1\,3\,6 \\ \times \quad 3 \\ \hline \end{array}$$

Two hundred and forty-seven

2
$$\begin{array}{r} 2\,4\,7 \\ \times \quad 3 \\ \hline \end{array}$$

One hundred and ninety-two

3
$$\begin{array}{r} 1\,9\,2 \\ \times \quad 3 \\ \hline \end{array}$$

Two hundred and six

4
$$\begin{array}{r} 2\,0\,6 \\ \times \quad 3 \\ \hline \end{array}$$

This is how I did 143 times 3.

You got the same answer as me!

1 4 3 times 3 is **1 0 0** times 3

add **4 0** times 3

add **3** times 3

300
1 2 0
9
4 2 9

Copy and complete.

5

1 0 0 × 4 = _____

3 0 × 4 = _____

so **1 3 0** × 4 = _____

6

$$\begin{array}{r} 130 \\ \times\quad 4 \\ \hline \end{array}$$

7

8 0 × 5 = _____

6 × 5 = _____

so **8 6** × 5 = _____

8

$$\begin{array}{r} 86 \\ \times\quad 5 \\ \hline \end{array}$$

9

2 0 0 × 3 = _____

5 0 × 3 = _____

so **2 5 0** × 3 = _____

10

$$\begin{array}{r} 250 \\ \times\quad 3 \\ \hline \end{array}$$

S

Try our 'halves' quiz. Work with a friend if you want to.

There are half a dozen questions!

1 How many minutes in half an hour?

2 How many months in a year and a half?

3 What is half of £1?

4 How many is half a dozen?

5 How many quarters are the same as $\frac{1}{2}$?

6 How many tenths are the same as $\frac{1}{2}$?

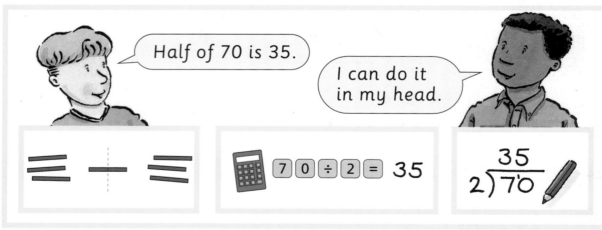

Half of 70 is 35.

I can do it in my head.

$70 \div 2 = 35$

$$\begin{array}{r} 35 \\ 2\overline{)70} \end{array}$$

What is half of:

7 90?   **9** 300?   **11** 500?   **13** 730?

8 120?   **10** 370?   **12** 520?   **14** 492?

This hat is half-price.

50 per cent off.

I'm not surprised!

50% is the same as <u>half</u> of something.

50% of £24 is £12

What is 50% of:

15 £15? **17** £3? **19** £57? **21** £17·50?

16 90p? **18** £3·60? **20** £185? **22** £9·50?

'50 per cent' means '50 out of every 100'.

23 What is 50% of £100?

24 What is 50% of £200?

25 What is 50% of £300?

26

I paid £27 for my coat in a sale. That was 50% off.

What was the price <u>before</u> the sale?

The three of us are renting a caravan.

It's £360 altogether.
How much is that, per person?

£360 ÷ 3

That's £120 per person, if 3 people go.

Use notes and coins to help you.

1 What if 4 people go ...
How much per person?

2 What if 5 people go ...
How much per person?

£360 ÷ 4

Seaside Campsite!
Luxury Rent-a-Tent
(maximum 4 people)

£84

3 How much per person if 2 people go?

4 How much per person if 3 people go?

5 How much per person if 4 people go?

Holiday cottage £700
(maximum 5 people)

6 How much per person if 2 people go?

7 How much per person if 3 people go?

8 How much per person if 4 people go?

9 How much per person if 5 people go?

We're going camping. We are not sure if our friend can come.

If 2 of us go, it will cost £13.50 per person.

If 3 of us go, it will cost £9 per person.

10 How much will it cost altogether?

Woodland chalet £645
(maximum 6 people)

11 How much per person if 2 people go?

12 How much per person if 3 people go?

13 How much per person if 4 people go?

14 How much per person if 5 people go?

≈ Take three numbers ≈

A game for 2 or 3 people.

You need: a tub of number tiles.

Take three numbers from the tub ...

and make the biggest number you can.

Then your friends do the same.

Six hundred and forty-three.

I win!

The person with the biggest number wins all these tiles.

Keep going until the tub is empty.

96   **Using understanding of place value**